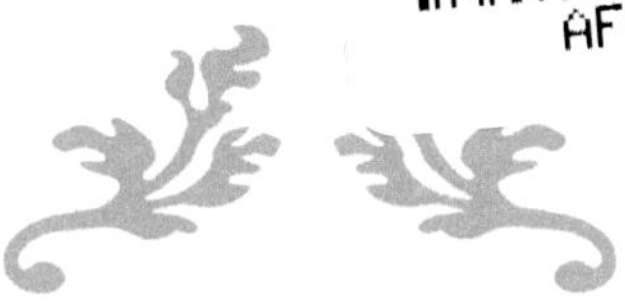

SYNDROMIC APPROACH TO MEDICAL DIAGNOSIS

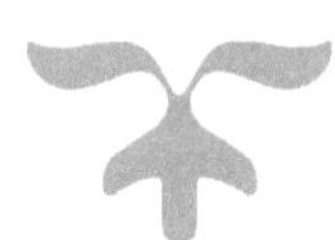

SUJIT K. BHATTACHARYA

SUDESHNA BARUA

ANISH BANERJEE

pencil

A brand of
One Point Six Technologies Pvt. Ltd.
123, Building J2, Shram Seva Premises,
Wadala Truck Terminal, Wadala (E)
Mumbai 400022, Maharashtra, INDIA
E connect@thepencilapp.com
W www.thepencilapp.com

PREFACE

Diagnosis of a disease is the most important step for rationale management of any patient. Classically, after scrupulous history taking, the clinician, when required, orders required tests to confirm the diagnosis. This traditional approach is highly suitable for physicians trained in modern medicine. However, an alternative approach of classifying patients according to certain presenting symptoms or on initial laboratory work up, will be helpful for beginners and for students and nurses to make the diagnosis which is called the syndromic approach.

This syndromic approach will provide the starting point to think of the aetiology of the syndrome or diseases. As for an example, a patient presents with breathlessness as the chief complaint, the clinician will be thinking breathlessness as a manifestation of certain diseases or a syndrome. This conceptual framework will guide the clinician to order further work-up including laboratory tests. This approach will be easy to follow and will minimise the tests required and cost of treatment.

This concise and handy book is dedicated for students beginning their clinical training, the nurses and all others involved in patient care.

— Sujit K. Bhattacharya
— Dr Anish Banerjee
— Sudeshna Barua

PULSE

When blood flows through the arteries, it is felt as rhythmical throbbing sensation. Pule is felt against a bony or firm surface for complete 1 minute. The principle is that the faster the pulse, you countless time. Radial pulse is most often used to examine the pulse and calculate the rate of the heart beats. If the pulse is not palpable due to hypotension or congenital anomaly, look for other accessible pulses, e.g., carotids, brachial, popliteal. It should be remembered that the carotids (both side) must not be palpated at the same time as it may stop circulation of blood into the brain. Sinus tachycardia, bradycardia, arrythmia, ectopics, atrial fibrillation (irregularly irregular), water hammer pulse (aortic incompetence and severe anaemia), feeble pulse (hypotension due to loss of fluid/blood/shock, sepsis) may be the detected if pulse is palpated carefully to detect not only the rate, but it can give a lot other information suitable for a syndromic approach to diagnosis. Abnormal pulsation may be due to venocculosive disease known as aortoaortitis (Takayasu's disease).

RATE CHANGES

Sinus Tachycardia (>100 bpm)	Sinus Bradycardia (< 60 bpm)
Anxiety	Athletes
Exercise	Meningitis
Thyrotoxicosis	Biliary cirrhosis
Excess Thyroxine supplementation	Typhoid fever (relative bradycardia)
Salbutamol	Complete heart block
Fever	β-Blockers
Adrenaline	Calcium channel blocker
Anaemia	Pontine haemorrhage
Acute blood loss	

PALPITATION

Palpitation means consciousness of the action of the beating heart.

Causes:
- Anxiety
- Congestive cardiac failure
- Myocardial infarction
- Thyrotoxicosis
- β-agnostics (adrenaline, salbutamol)

Symptomatic treatment of palpitation: Rest, Propranolol, Anxiolytic

BLOOD PRESSURE (BP)

BP is defined as the lateral pressure exerted by the flowing blood within the arterioles. BP should always be checked at least twice with an interval of 5-10 minutes in the same sitting.

BP: Normal: less than 120/80 mmhg in sitting position
BP: Pre-hypertension: 120 to 139 mmhg
BP: Hypertension: > 140/90 mmhg

Essential hypertension is by exclusion of secondary cause.

Secondary hypertension means when there is a cause for hypertension:

- Thyrotoxicosis
- Hypothyroidism
- Pheochromocytoma
- Renal artery stenosis
- Acromegaly
- Gigantism

- Conn's syndrome (Primary hyperaldosteronism)
- Cushing's syndrome

Note*: Hypertension may co-exist with Diabetes mellitus, lipid abnormalities and obesity.*

Treatment:

Maintaining body weight according to height, weight and BMI

Exercise: Walking daily for 30 minutes 5 days in a week

Basal Metabolic Rate (BMR):
- Normal: 18.5-24.9
- Over weight: 25-29.9
- Obese: 30-39.9

Metabolic syndrome:
- Diabetes mellitus type 2 (Insulin resistance)
- Abdominal obesity
- Dyslipidaemia

Drug Treatment:
- Diuretics: Hydrochlorothiazide (12.5 mg)
- Calcium Channel blockers: Amlodipine, Diltiazem
- Angiotensin converting enzyme (ACE) inhibitor: ramipril, lisinopril
- Angiotensin receptor blocking (ARB) agent: Olmesartan (20-40), Telmisartan (40-80)

- β-blockers: Atenolol (25-50 mg), Metoprolol (25-50 mg), bisoprolol, labetalol, carvedilol
- α_1-blocker:Prazocin (5mg),
- Hydralazine, methyl dopa,

Target organ damage by longstanding hypertension:
- Heart—Left ventricular hypertrophy (LVH) and failure (LVF)
- Kidney- chronic kidney disease (CKD)
- Retina: Hypertensive retinopathy
  Brain vessels: Intracerebral haemorrhage (ICH), Subarachnoid haemorrhage (SAH -Ruptured Berry's aneurysm)

HYPOTENSION:

Systolic BP < 90 mmhg

Treatment:
Symptomatic: Correction of hypovolemia by Normal saline (NS), Ringer's lactate (RL), ---

Noradrenaline, Dopamine, Dobutamine, Adrenaline

THERMOREGULATION

- Normal body temperature is maintained and regulated by the hypothalamus, skin, sweat glands and circulatory system in human.

Temperature
- Recorded with a thermometer in Degree Centigrade or Fahrenheit.
- Temperature is recorded in the axilla, under surface of the tongue, or in the rectum, particularly in children; aural and skin temperature is also recorded. Axillary temperature is not so accurate and should be avoided as far as possible.
- Normal oral temperature: 98.6-degree F;
- Fever > 98.6-degree F (oral);

Types of fever:
- Intermittent: fever which has a fluctuating base-line
- Remittent: fever fluctuates, comes and goes, but never becomes normal

- Continuous: temperature is prolonged (days), but never much variation
- Relapsing: spikes of fever for days with remissions for some days to reappear again
- Hectic: fever where temperature range varies widely

Causes:
- Infections: bacteria, viruses, rickettsia, protozoa
- Non-infectious diseases
- Malignancy
- Drugs
- Unknown

Some causes of fever:

Viral fevers, Typhoid, Urinary tract infection, Hepatitis, Tuberculosis, Respiratory Tract Infections, Acute diarrhoea/dysentery, Boils, Skin infections, Tonsillitis, Mumps, Measles, Dengue, Malaria, Kala-azar, Typhus, Sore throat, Ear infection, HIV infection

Symptomatic Treatment of fever: Sponging, Paracetamol, Ibuprofen, Diclofenac .

Targeted Treatment: Chloroquine for Malaria, Antibiotic for Pneumonia, Aspirin for Rheumatic fever, Steroid for Immunological diseases

Hypothermia

Hypothermia is defined as a body temperature below 95-degree F or 35-degree centigrade. This is a grave medical emergency. Hypothermia is caused by exposure to extreme cold or emersion to cold water. The heart, brain and other vital organs cannot function properly and the person dies. Treatment is to worm up the person. The patient should be handled gently. The patient should be taken out of cold exposure, wet clothes removed and covered with warm blankets. The patient should be given warm beverages. Warm intravenous fluid may be administered. Direct heat should be applied to the body particularly the extremities.

Hyperpyrexia:

Hyperpyrexia is defined as when the body temperature rises above 106.7-degree F or 41.5-degree C. This condition can be caused by infections, e.g., Malaria, Viral infections. Treatment is primarily directed to the cause. Temporarily high temperature can be lowered by administering paracetamol, NSAIDs, and non-pharmacological method, e.g., tepid sponging (the body temperature being higher than air/water temperatures, tepid sponging will facilitate dissipating body heat to move out of body to outside air molecules being in direct contact with the skin).

Heat stroke:

When the body temperature rises above 104-degree F, it is called heat or sun stroke. Exposure to hot atmosphere commonly caused heat stroke due to failure of the thermoregulation mechanism of the body. Besides high body temperature the patient may be dehydrated with absence of sweating and accompanied by vomiting, seizures, muscle cramps and muscle weakness and shallow breathing. Ultimately, vital organs fail to function. Treatment is to cooling of the body using cold water, fanning, air-conditioner and ice in the arm pits (axilla), groins, elbows as they are rich in blood circulation. Even the patient may be emersed in ice-cold water to dissipate the body heat. Intravenous fluid may be required. This condition should be considered a serious medical emergency.

Fever in ICU:

Once the patient is in the ICU for a few days, it is most likely that the fever has been acquired in the ICU. Persistent rise of temperature for a few days is most likely indicative of fever than a single spike of temperature. Fever in ICU can be due to infections acquired in the ICU or due to non-infectious causes.

Infectious causes are:
- Ventilator-associated Pneumonia
- Catheter-related Urinary Tract Infections
- Infected Bed Sore
- Infection of surgical wounds
- Pseudomembranous colitis (Clostridium difficile infection)
- Surgical wound-related infection

High lactate level (> 2 mmol/liter) is usually seen in sepsis. Lactate levels are high because of increased lactate production due to anaerobic metabolism and decreased renal clearance. C-reactive protein (CRP) is an acute-phase reactant and Procalcitonin is an indicator (biomarker) of severity of bacterial infection. Blood count, kidney and liver function tests, including serum amylase and lipase are required (if pancreatitis is suspected).

Fever with chill and rigor
- Malaria
- Urinary Tract Infection
- Cholangitis
- Abscess

Malaria

Malaria is a parasitic disease mostly seen in the tropics and sub-tropical regions. Malaria is caused by *Plasmodium Vivax, Plasmodium Falciparum, Plasmodium Malariae* and *Plasmodium Ovale.* The disease manifests with fever accompanied by chills and rigor and the patient may also vomit. The temperature may go up to 104-105 degree. Liver and particularly spleen is enlarged. Diagnosis is made by examination of thick and thin blood smear for demonstration of the malaria parasite. The positive results (demonstration of the parasite) is more often detected when smears are taken at the febrile stage. A negative test does not exclude malaria. Repeated examinations may be required. Rapid diagnosis by testing for antigens common to all Plasmodium spp. is available. Vivax malaria is treated with chloroquine in areas where the parasites are sensitive to the drug. The dose is Chloroquine base 600 mg orally in full stomach (to reduce the chance of vomiting) followed by 300 mg base after 6 hours and 300 mg base for 2 days. Falciparum malaria may be drug resistant and often to chloroquine. Oral artemisinins are effective in the treatment of uncomplicated falciparum malaria, although recrudescence is high. FALCIGOPLUS kit (artesunate 50 mg tab + mefloquine 250 mg tab) and artemether (80 mg twice daily and lumefantrine 480 mg twice daily) for three days is also available. Sulfadoxine (500 mg)-pyrimethamine (25 mg tab) three tablets as single dose is effective for the treatment of falciparum malaria. For radical cure of vivax malaria, primaquine (15 mg) daily for two weeks is used along with full dose of chloroquine. Quinine sulphate has been used for the treatment of malaria.

Typhoid fever

Typhoid fever (enteric fever) is caused by *Salmonella Typhi* and transmitted by ingestion of facially contaminated food and water. The disease is charactered by fever which slowly increases and reaches to about 104-degree F in untreated cases and lethargy, weakness and occasionally vomiting particularly in children. Patients may have cough. Several serious complications have been reported in association with typhoid fever, e.g., intestinal haemorrhage and perforation. Diagnosis is made clinically by prolonged fever, leukopenia and blood culture. Typically, blood culture is positive in the first week of the illness. Later on, stool and urine cultures may be positive. Serological test, e.g., Widal test is non-specific and cannot be relied upon to make a diagnosis. Typhi dot serological test can be done but specificity and sensitivity in not high. Treatment is rest, antibiotic and feeding. Chloramphenicol was the drug of choice for long time, but chloramphenicol resistant *salmonella typhi* have been reported. The other drugs effective in the treatment of typhoid fever are co-trimoxazole and amoxicillin that also became ineffective. Fortunately, ciprofloxacin was marketed and was very effective and given orally. Ceftriaxone IV and azithromycin are effective drugs for the treatment of the typhoid fever. Relapses have been reported and should be treated in the same way for 14 days. Typhoid carriers have been reported. Cholecystectomy may be required. Three typhoid vaccines are available, e.g., Ty21a (a live vaccine given orally),Typhoid conjugate vaccine (TCV) and an injectable Vi- capsular polysaccharide (given IM) vaccines.

Urinary tract related symptoms

Increased frequency: Urinary tract infection (UTI), diabetes mellitus and insipidus, enlargement of prostate, excessive drinking

Blood in urine: Haematuria
Lymph in urine: Chyluria
Burning in urine: UTI

Routine urine examination showing pus cells > 5 hpf should be cultured for identification of the incriminating organism and sensitivity testing. Nitrofurantoin may be empirically prescribed pending the culture and sensitivity test report. Recurrent UTI may occur particularly in women that should investigated for any obstruction in the urinary tract. An Ultrasound KUB is useful.

Retention of urine-enlargement of prostate, cerebrovascular accidents, medications (antispasmodics, anticholinergics, antidepressants – amitriptyline having anticholinergic component), stone in the urethra, stricture (post-gonococcal infection- Gonorrhoea), surgery. Neurogenic bladder is due to disease of the nerves supplying the bladder-spinal cord injury, multiple sclerosis, parkinsonism, stroke, spina bifida etc.

RASH

Many diseases present with fever and rash. These rashes have a set pattern which helps in clinical diagnosis of the disease. Some of the common diseases associated with fever and rash are:

Measles	Rubella
Small pox	Chicken pox
Dengue	Herpes Zoster
Syphilis	Chicken pox
Steven-Johnson Syndrome	Shigellosis
Toxic Ectodermal Necrosis - (TFN)	Typhoid Fever
Chikungunya	HIV
Idiopathic Thrombocytopenic Purpura (ITP)	
Drug rash	Leishmaniasis
	Others
Ebola	

ANAEMIA

Anaemia is defined as reduction of concentration of haemo-globin in blood. Anaemia is present when Hb < 14g/dl (men) and <12g/dl (female). Anaemia can be acute or chronic.

Acute anaemia is caused by acute loss of blood:

Bleeding

- Trauma
- Hematemesis
- Melena
- Ruptured oesophageal varix
- Haemoptysis
- Epistaxis
- Piles

Symptoms: (Depends upon amount of blood lost. When Hb <7g/dl, it becomes symtomatic)

- Thready pulse
- Low BP (Hypotension)
- Cold periphery
- Palpitation

- Breathlessness
- Lethargy
- Syncope
- Cardiac pain

Diagnosis:
- Evidence of blood loss
- Clinical features
- Packed cell volume
- Underlying disease

Management:

Blood loss must be replaced by blood. Grouping and cross matching is required before blood is transfused. Underlying disease identification and management to prevent further bleeding

Chronic anaemia:
- Iron deficiency anaemia: hypochromic microcytic, hypochromic normochromic:
- Vitamin B12, Folic acid or both deficiency, malabsorption - megaloblastic anaemia
- Drug Induced, slow oozing-Aspirin (for stroke prevention), NSAIDs, Chloramphenicol,
- Leukaemia
- Aplastic or hypoplastic anaemia
- Multiple myeloma

- Chronic Kidney Disease
- Chronic infections-Tuberculosis, HIV/AIDS
- Hemoglanopathy -Thalassemia, Sickle Cell Disease
- Cancer
- G-6PD-deficiency
- Haemolytic anaemia
- Autoimmune Haemolytic Anaemia

Chronic anaemia is better tolerated than acute anaemia. Pallor, weakness, lethargy, dizziness, chest pain on exertion, palpitation, weight loss.

Diagnosis:
- History of dietary habit
- Peripheral blood smear examination-size and shape of RBC, abnormal cells, stanning- malaria parasite, L.D bodies (intracellular)-Visceral Leishmaniasis
- Bone marrow examination- hypoplastic, immature RBC, leukaemia-blast cells, malaria parasite, leishmania donovani (LD bodies)
- B12, Folic acid estimation

Management:
- Balanced diet
- Oral Iron: Ferrous sulphate, Ferrous Gluconate, Ferrous chloride
- Intravenous: Iron-Dextran
- Folic acid, Vitamin C

- Injection Erythropoietin
- Blood Transfusion
- Bone marrow transplantation
- Stem Cell Therapy

Thalassemia

Thalassemia is life-long blood disorder which is inherited from the parents. This disease is a disorder of the haemoglobin which is present in the RBC. They carry oxygen to the tissues. The normal haemoglobin contains two alfa and two beta chains. Normally after birth the foetal haemoglobin disappears and adult haemoglobin is formed. Thalassemia is classified into alfa-thalassemia and beta-thalassemia. Thalassemia may also be classified into trait, minor and major depending upon the severity. As a result of abnormal haemoglobin, the cells prematurely breakdown liberating the haemoglobin. The haemoglobin gives rise to bilirubin and unconjugated bilirubin is formed. This is manifested by anaemia and jaundice and splenomegaly. Blood transfusion is required repeatedly to correct anaemia otherwise skeletal deformity occurs with growth retardation. Iron should not be given. Daily oral folic acid tablets are given. Bone marrow transplantation is curative.

CYANOSIS

Cyanosis is defined as bluish discolouration of mucous membra, skin, and nail beds. It is best detected in the tips of the nose, tongue and fingers and is due to excess of reduced haemoglobin in the arterial blood (> 5g/dl). Cyanosis can be detected clinically when the amount of reduced (deoxygenated) haemoglobin in the blood is at least 5 grams/L.

Types: Central, peripheral, mixed and differential cyanosis.

Central:

Congenital cyanotic heart disease

- Tetralogy of Fallot
- Transposition of great arteries
- Atrial or ventricular septal defects with reverse shunt (Eisenmenger complex) due to development of pulmonary hypertension
- Total anomalous pulmonary venous return

Lung diseases:
- Pneumonia
- Chronic Obstructive Pulmonary Disease (COPD)
- Respiratory failure
- Pulmonary oedema
- Pulmonary embolism

Peripheral cyanosis:
- Hypotension
- Hypovolemia

CLUBBING

Clubbing of fingers

Causes:

- Congenital
- Congenital Cyanotic Heart Diseases
- Bronchogenic Carcinoma
- Bronchiectasis
- Lung Abscess
- Bacterial endocarditis
- Eisenmenger Complex
- Biliary Cirrhosis
- Empyema

OEDEMA

Oedema means accumulation of excess fluid in the subcutaneous tissue.

Causes:
- Congestive Cardiac failure
- Cirrhosis of the Liver
- Malnutrition (hypoproteinaemia)
- Nephrotic Syndrome
- Acute Glomerulonephritis
- Deep Vein Thrombosis (DVT) -unilateral oedema
- Filariasis
- Hypothyroidism
- Liquorice
- Drugs- Amlodipine

Symptomatic treatment of oedema: Diuretics (Furosemides, Torsemide, Combination of furosemide and spironolactone). Treatment of the cause, e.g., cardiac (CCF) lung diseases (Cor-pulmonale) or hypoproteinaemia due to malnutrition / cirrhosis of the liver

SWELLING OF THE BODY

Generalised: Acute glomerulonephritis, nephrotic syndrome, congestive cardiac failure, cirrhosis of the liver, hypoproteinaemia (malnutrition); myxoedema (non-pitting),

Bilateral limb swelling (oedema): CCF, Other causes of generalised oedema, amlodipine

Unilateral oedema: Deep vein Thrombosis (DVT), diabetic foot, lymphangitis (Filariasis)

Ascites is collection of free fluid in the peritoneal cavity due to hypoproteinaemia and portal hypertension (cirrhosis liver)

Management: Treatment of the cause. Diuretics: furosemide, torsemide, spironolactone, combination of spironolactone and furosemide; give plenty of smashed egg white orally, 20% 100 ml human serum albumin by slow intravenous infusion

JAUNDICE

Jaundice is characterised by yellowish discolouration of mucous membrane, and skin. It is best seen in the upper and outer surface of the sclera, under surface of the tongue, and upper palate and in advanced cases skin all over the body. Jaundice is due to excess accumulation of bilirubin in the blood. When bilirubin concentration is > 3 mg/dl, jaundice can be detected clinically (examination must be done in natural sun light).

Bilirubin metabolism: when RBC is haemolysed, haemoglobin is liberated in the blood. Heme part of the haemoglobin is broken down into unconjugated bilirubin which is converted in the liver into conjugated bilirubin. This is excreted from the liver and concentrated in the gallbladder (bile). The bile is excreted into the intestine and converted into stercobilinogen by the colonic bacteria and finally excreted as stercobilin in the stool. A small part of the stercobilinogen is reabsorbed into the blood and excreted into the urine as urobilin. In all these stages of bilirubin metabolism, defect can cause hyperbilirubinemia.

Types:
Pre-hepatic:

Haemolytic anaemia:

- Thalassemia
- G-6PD deficiency
- Auto-immune haemolytic anaemia

Hepatic (intrinsic):

- Infective hepatitis
- Cirrhosis of the Liver
- Hepatic carcinoma
- Alcoholic liver disease
- Hemochromatosis
- Drug induced liver disease-
    - aflatoxin
    - pyrazinamide
    - INH
    - rifampicin

Post-hepatic:

Obstruction to the common bile duct:
    - Gall Stone
    - Carcinoma of the Head of the Pancreas

Classification based on conjugation of bilirubin:

Unconjugated hyperbilirubinemia
    - Gilbert syndrome
    - Crigler-Najjar syndrome

Conjugated hyperbilirubinemia
- Dubin Johnson
- Rotor

Ascites

Ascites is characterised by accumulation of excess free fluid in the peritoneal cavity. Only in female there is normally very little free fluid. Ascites can be detected clinically when the amount of free fluid exceeds 1500 ml. The umbilicus is everted, shifting dullness and fluid thrill may be elicited. Abdominal veins may be seen (caput medusa). Ultrasonography can detect free fluid (ascites) in the peritoneal cavity and enlargements of liver and spleen can be detected. Splenomegaly and portal hypertension cause ascites.

Aetiology:
- Cirrhosis of the liver
- Congestive cardiac failure
- Intestinal tuberculosis
- Malignant growth of different abdominal organs (liver carcinoma, carcinoma of the female reproductive organs, carcinoma of the stomach, pancreas and others) and lymphomas.
- Malnutrition (low serum albumin)

Diagnosis:
- History / Clinical examination

- Examination of aspirated ascitic fluid (diagnostic tap)
  - Colour
  - Cell count/cell type
  - Protein

Ascitic fluid are of two types:

Transudate: Ascitic fluid Protein < 2.5g/dl & cell count < 500/dl

Exudate: Ascitic fluid Protein >2.5 g/dl & cell count >5000/dl

SAAG (Serum-ascites albumin gradient) is > 1.1 in cases ascites is due to portal hypertension. In tubercular ascites, ascitic fluid protein > 2.5 g/dl (exudate) and cells are pre-dominantly lymphocytes. CT-scan and Doppler studies are required in selected cases.

Treatment;

- Treatment of the cause
- Diuretics (furosemide, spironolactone or combination)
- Therapeutic Tapping: Excessive accumulation of ascitic fluid causes distension of the abdomen and pushes the diaphragm towards the chest resulting in abnormally more pressure on the heart and lungs

causing difficulty in breathing. In such situations, not more than 1500 ml of fluid should be aspirated in one seating. Tapping of ascitic fluid may precipitate hepatic coma in cases of cirrhosis of the liver.

HAEMOPTYSIS

Haemoptysis is coughing out of fresh blood. It is a frightening symptom.

Causes:
- Tuberculosis
- Bronchogenic carcinoma
- Pneumonia
- Lung infections
- Pulmonary embolism
- Blood diseases
- Foreign body in the respiratory tract/lungs

Treatment:
- IV Fluid
- Blood transfusion
- Antibiotic/antitubercular drug

HEMATEMESIS

Hematemesis means vomiting out blood.

Causes:
- Bleeding peptic ulcer
- Ruptured oesophageal varix
- Trauma
- NSAIDs, Aspirin, Steroid
- Carcinoma of the stomach

MELENA

Melena means passing out of stools mixed with altered tarry blood.

Causes:
- Bleeding peptic ulcer
- Aspirin for prophylaxis for prevention of thrombosis

Haematochezia should be differentiated from melena. In haematochezia fresh blood is passed through the anus.

BLEEDING PER RECTUM

- Haemorrhoids
- Rectal polyp

EPISTAXIS

Epistaxis is defined as bleeding(fresh blood) from the nose. Sometimes the blood loss enough to produce shock.

Causes: I
- Infection and inflammation localised inside the nose .
- Hypertension
- Blood disorders
- Trauma
- Foreign body

And in some cases, the cause in uncertain.

Investigations: TC,DC. Hb%, ESR, Peripheral blood smear examination. Swab obtained from the nose -Gram staining and culture

Management: Nasal pack with small amount of adrenaline

If hypertensive, lower the BP
Treat infection-use amoxicillin
ENT Referral

HAEMATURIA

Haematuria is defined as passage of blood in urine.

Causes:
- Acute glomerulonephritis
- Urinary tract infection
- Enlarged prostate
- Tumour in the Urinary Tract
- Bleeding diathesis
- Trauma
- Urinary stone
- Carcinoma of the Kidney
- Polycystic kidney disease

SYNCOPE

Syncope is transient loss of consciousness for a few seconds without residual neurological deficit upon recovery. Vasovagal syncope is caused by pooling of blood in the legs during prolonged standing. As a result, the blood supply to the brain become less and the person faints and falls on the ground with return of consciousness. There is no residual neurological change.

Micturition syncope occurs, mostly in elderly persons, during or immediately after micturition due to drop in blood pressure, although the exact cause in not clear. This generally occurs when the patient gets up to urinate from deep sleep at night. This type of syncope can be avoided by sitting during urination. Covid-19 patients have been reported to develop micturition syncope. Other types of syncope are situational syncope (dehydration, anxiety, hyperventilation), postural syncope, cardiac syncope (aortic stenosis)

VERTIGO

Vertigo is relatively common experience. It means spinning of the head or surrounding. Vertigo can be caused by Vestibular neuritis or labyrinthitis, Meniere's disease (vertigo, tinnitus and hearing loss) and positional vertigo. Treatment is to identify the cause and treat. Consultation with ENT specialist is advised. Low salt diet and Cinnarizine may give relief.

CONVULSION

Convulsion is defined as sudden onset of repeated tonic, clonic movements of the limbs with involuntary urination and tongue bite.

Causes:
- Febrile convulsion (children)
- Epilepsy
- Hypoglycaemia
- Brain tumour
- Stokes-Adam syndrome
- Poisoning (Strychnine, Organophosphorus)

Treatment:
- Airway:
- Prevent tongue bite
- Lorazepam 4 mg IV slowly, may be repeated
- Blood sugar, ECG, CT-Brain, EEG
- Refer to Neurologist

COMA

Coma is a state of prolonged unconsciousness when the patient cannot be aroused even by deep painful stimulus. Coma is differentiated from Brain death where the brain has undergone irreversible damage but the heart may continue to beat. Although the patient may regain consciousness from coma, but in brain death, the patient cannot survive. Coma may lead to brain death.

Causes of coma:
- Epilepsy
- Cerebral Haemorrhage
- Hypoglycaemia
- Diabetic Ketoacidosis
- Hysteria (functional)
- Over dose of narcotics/sedatives
- Electrolyte imbalances

Glasgo-Coma Scale: This scale guides the clinicians to monitor status and changes of the level of consciousness with regard to time and with or without active interventions.

Score	Eye Opening	Verbal Response	Motor Response
1	None	None	None
2	To pain	Incomplete sounds	Abnormal extension
3	To voice	Inappropriate words	Abnormal flexion
4	Spontaneously	Confused	Withdrawal from pain
5		Oriented	Localizes to pain
6			Obeys commands

Maximum score:15; Head Injury-Severe: 8 or less; Moderate: 9-12; Mild: 13-15; Coma: 3-8.

Brain death is defined as a condition when an unconscious patient cannot be aroused by external stimuli, although the heart may continue to beat and respiration can be maintained by ventilation. Such a patient cannot regain consciousness and will die.

ACUTE DIARRHOEA

Acute diarrhoea is defined as passage of 3 or more loose stools in 24 hours. When loose stools contain blood, it is known as Dysentery. Dysentery is accompanied by fever and abdominal pain and tenesmus (sense of incomplete defecation with strangury).

Aetiology:

Bacteria:
- *Vibrio cholerae* O1 & O139
- *Eschesheria coli*
    - Enterotoxigenic
    - Enteropathogenic
    - Enterohemorrhagic
    - Enteroadherent
- *Vibrio parahaemolyticus*
- *Non-typhoid salmonellae*
- *Clostridium difficile*
- *Campylobacter jejuni*

Viruses:

Rotavirus

Parasites:

- *Giardia lamblia*
- *Cryptosporidium*

Management:

Mild and moderate cases: Oral Rehydration therapy using Oral Salt solution recommended by WHO. A hypoosmolar ORS has also been recommended. The compositions of both are as below:

Composition (mml/L)	Standard WHO-ORS	Low Osmolarity ORS
Glucose	111	75
Sodium	90	75
Potassium	20	20
Chloride	80	65
Citrate	10	10
Total Osmolarity	311	245

Table: Composition of ORS

SHIGELLOSIS

Shigellosis is characterised by frequent passage of loose stools mixed with blood and mucous, fever and accompanied by abdominal cramps and tenesmus (incomplete sense of defaecation). *Shigellae* spp. are: *Shigella dysenteriae, S. sonni, S.boydii* and *S. flexneri.* (Shigellosis). The other causes of acute bloody diarrhoea are:

- *Entamoeba histolytica.*
- Enteroinvasive *E. coli*

Shigellosis caused by S. dysenteriae type 1 is multi-drug resistant and possess a therapeutic challenge. Treatment of shigellosis requires antibiotics, e.g., Ciprofloxacin 500 mg twice daily x 3-5 days. Oral antibiotic is effective and preferred. Ceftriaxone is also effective but needs to be given by intravenous injections. Complications of shigellosis include rectal prolapse, febrile convulsions in children, arthritis, toxic megacolon and Haemolytic-Uraemic syndrome (HUS). HUS is a triad of haemolytic anaemia, kidney failure and thrombocytopenia.

Rotavirus diarrhoea is common in children aged between 6 months to 2 years. Early vomiting is an important sign of rotavirus diarrhoea. Rotavirus vaccines are available.

Cholera is caused by *Vibrio cholerae* O1 and O139. Cholera is characterised by frequent passage of loose stools (rice-water stools) having a fishy smell, vomiting and life-threatening dehydration. Asymptomatic infections and mild cases also occur. Management of cholera is essentially correction of dehydration and maintenance of hydration by Oral Rehydration Salt solution recommended by World Health Organization. Intravenous fluid (Normal saline or Ringer's lactate) is required for severely dehydrated cholera cases. Antibiotics (tetracycline or single dose doxycycline, ciprofloxacin, norfloxacin) are given along with rehydration fluids. Antibiotic treatment reduces the stool volume, minimises the fluid requirement, cuts down the duration of hospitalization and hastens recovery. Antiemetics, charcoal, kaolin and pectin are not recommended. Effective Oral cholera Vaccine is available.

Cryptosporidium causes chronic diarrhoea particularly in immunocompromised people.

Clinical feature:
- Loose watery stools; Dehydration (mild, moderate, severe)
- Frequently

- Vomiting
- Abdominal pain (absent in cholera)
- Fever (absent in cholera)

Management:

- Rehydration: Oral using Oral Rehydration Salt solution
- Intravenous using Normal saline or Ringer's lactate
- Antibiotic for cholera only (Tetracycline, Doxycycline single dose, Ciprofloxacin)
- Feeding
- Zinc supplementation

Prevention:

- Improving sanitation
- Safe drinking water
- Handwashing
- Vaccination: Oral Heat-killed cholera Vaccine: Protection 66% of vaccinees for 5 years

CONSTIPATION

Constipation means infrequent passing of stools which gives a feeling to the person that stools have not been passed enough per week. Normal individuals pass stools once or twice daily or weekly and it varies. The stools are hard, because of the long transit time, much of the water from the stool is absorbed in the colon. Causes: drink too little water, non-fibrous food, no exercise or movement, irritable bowel syndrome, hypothyroidism, anticholinergic drugs, antidepressants with anticholinergic property. Relief may be obtained by drinking lots of water, fibrous food, exercise and other medication for the specific condition.

SWELLING OF THE ABDOMEN

- Fat
- Fluid
- Pregnancy
- Ovarian cyst

ABDOMINAL PAIN

Upper abdomen

Gallbladder:
- Cholecystitis
- Gall stone
- Cholangitis
- Peptic ulcer
- Pancreatitis
- Splenic infarct & rupture
- Trauma

Lower abdomen:
- Renal Colic
- Kidney stone
- Ureteric stone
- Urinary Bladder stone
- Prostatitis (male)
- Retention of urine
- Oophoritis (female)

- Salpingitis (female)
- Ruptured ectopic pregnancy (female)
- Appendicitis
- Intestinal colic
- Peritonitis
- Colitis
- Trauma

History taking and Clinical examination is of utmost importance to diagnose the cause of abdominal pain. Ultrasonography (USG) is the single most important investigation to ascertain many of the causes of abdominal pain.

CHEST PAIN

Heart:

- Acute Coronary Syndrome:
  - ST-segment Elevation Myocardial Infarction (STEMI)
  - ST-segment Depression Myocardial Infarction (NSTEMI)
  - Angina pectoris
- Pericarditis
- Pericardial effusion

Lungs:

- Pleuritis
- Pleural effusion
- Pneumonia
- Pneumothorax
- Pulmonary embolism
- Penetrating injury

Chest wall:

Herpes zoster (Singles)- lesions of one side does not cross the mid-line to the opposite side-distribution of lesions along the nerves, e.g., intercostal nerves, ophthalmic division of the Trigeminal nerve etc.

Myocardial Infarction

Characteristics of cardiac chest pain: severe, excruating or oppressing or squeeze pain in the chest associated with profuse sweating and the pain may radiate to the arms or throat with profuse sweating and hypotension.

STEMI/ NSTEMI

Hospitalization

ECG, Echocardiography, CBG, Lipid profile, Thyroid profile, Blood counts, CRP

Cardiac biomarkers: CPK, CPK-MB, Troponin I & T

Stress Testing

Coronary angiography

Diet: Nil by mouth for 6 hours

Oxygen, Morphine, Isosorbide dinitrate under tongue, Aspirin, Clopidogrel, Atorvastatin, β-blocker, Ace Inhibitor

Treat Hypertension and lipid abnormalities, if present

Treat hypotension, if present

Thrombolysis (best if the patient presents early, not more than after 6 hours) Except NSTEMI & Unstable angina

PTCA, Coronary angiography, Coronary Artery Bypass Grafting (CBGA)

ARRYTHMIA

Normal QRS Complex (0.11 second)
Tachyarrhythmia

Narrow complex QRS (<0.11 sec)
Atrial fibrillation (AF)
Atrial flutter
Supraventricular tachycardia

Wide complex QRS (>0. 11 sec)
Ventricular tachycardia (VT)
Ventricular Fibrillation (VF)
(SVT)

Atrial fibrillation:

Causes:

- Post-myocardial infarction
- Coronary artery disease
- Thyrotoxicosis
- Mitral stenosis
- Cardiomyopathy
- Hypertension
- Pericarditis

- Elderly people
- Adrenaline
- Pheochromocytoma
- Lone AF (AF with no known cause)

Diagnosis: Pulse-irregularly irregular, ECG-P-waves are replaced by fibrillary f waves with irregular rhythm.

Treatment: Calcium channel blockers, β-blockers (metoprolol), digoxin or amiodarone. Cardioversion is highly effective.

Complication: Pulmonary or systemic embolism. Anticoagulants e.g., heparin, low molecular heparin (LMWH) or warfarin or oral anticoagulant, e.g., apixaban (2.5 to 5 mg daily) are given to prevent thromboembolism.

SVT: intravenous adenosine, β-blocker

Ventricular tachycardia and ventricular fibrillations are life-threatening conditions and best treated by defibrillation.

ELECTROLYTES

Sodium (Na⁺):

- Normal : 135-141 mEq/L
- Hyponatremia: Sodium < 135 mEq/L
- Hypernatremia: Sodium > 141 mEq/L

Potassium:

- Normal : 3.5 -5-mEq/L
- Hypopotassaemia: < 3.5 mEq/L
- Hyperpotassaemia: > 5.0 mEq/L

Calcium:

- Normal: 8.5-10.3 mg/dl
- Hypocalcaemia:< 8.4 mg/dl
- Hypercalcemia: > 10.3 mg/dl

Magnesium:

- Normal: 1.5 mg/dl
- Hypomagnesemia: < 1.3 mg/dl
- Hypermagnesemia: > 2.6 mg/dl

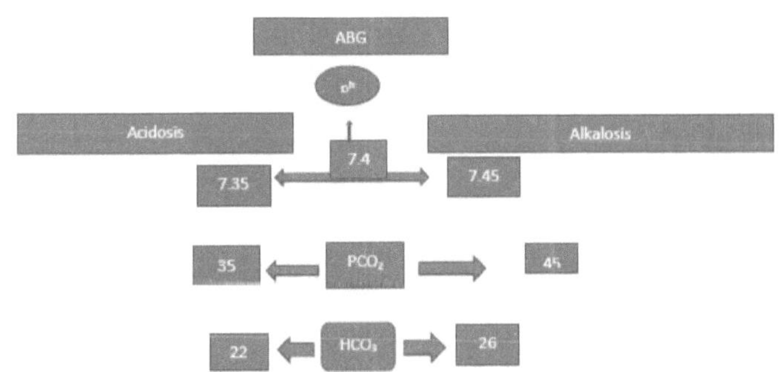

Acid-Base Balance

Respiratory Acidosis		↑	Metabolic Acidosis		↓
	PCO2			HCO3	
Respiratory Alkalosis		↓	Metabolic Alkalosis		↑

CONVULSION

Convulsion, commonly known as fits, is a tonic, clonic movements of the limbs accompanied by tongue bit and involuntary urination. the patient may become unconscious.

Causes:
- Febrile convulsion in children
- Epilepsy
- Hypoglycaemia
- Electrolyte abnormalities
- Meningitis/ Raised intracranial pressure
- Stokes-Adam attack
- Brain tumour
- Strychnine / organophosphorus poisoning
- Hysteria (functional)

Management: Ensure airway (A), breathing (B) and circulation (C); prevent tongue bite, place the patient in lateral position and injury to the limbs; give lorazepam 4 mg by slow IV and maybe repeated after 5 minutes, if convulsion not controlled. Second line drugs are phenytoin, phosphophe-nytoin, Levetiracetam. If not controlled, consultation with neurologist / anaesthesiologist.

Hypoglycaemia

Hypoglycaemia should be recognised in patients receiving antidiabetic medicines (oral drugs/ insulin) missed meal or undertaken rigorous exercise. The patient is observed to sweat profusely with palpitation, tremor and blurring of vision, fits and come. The blood sugar level is usually below 72 mg/dl. If the patient is conscious, give glucose/sugar immediately by mouth till the symptoms are gone and blood sugar level goes above 90 mg/dl. If Hypoglycaemia is severe, give intravenous glucose 25% 50 ml rapidly. If required, it may be repeated. Keep in mind that hypoglycaemia may appear again and should be treated promptly. All patients receiving antidiabetic medicines should be educated about the symptoms of fall of blood sugar and what to do in such situation. The patient should keep sugar handy at home and carry sugar whenever going out.

Hyperglycaemia

Hyperglycaemia is caused by lack or reduced production or resistance of insulin elaborated from the islet cells of the pancreas. This is known as **Diabetes Mellitus.** Diabetes is characterised by hyperglycaemia and the classical symptoms are polyuria, polydipsia and polyphagia.

Diagnosis: Fasting blood sugar > 126 mg/dl or 2-hour post-prandial blood sugar > 200 mg/dl. Fasting blood sugar 100-125 mg/dl is called prediabetes and 145-199 is called impaired glucose tolerance. HBA1c (glycated haemoglobin) measures the average blood sugar level during the last 3 months. The target HBA1c is < 7.0%. It indicates the control of blood sugar. Sometimes hyperglycaemia is for the first time encountered in pregnancy, which may or may not disappear after pregnancy (Gestational diabetes). As opposed to diabetes mellitus, diabetes insipidus is charactered by polyuria due to deficiency of antidiuretic hormone secreted from the posterior pituitary gland in the absence of hyperglycaemia. Diabetes mellitus is classified as below: type 1 which is seen in individuals below 40 years of age (previously called juvenile diabetes) and either have lack of insulin production or insulin resistance. The type 2 diabetes generally seen in those above the age of 40 years (previously called maturity onset diabetes) is characterised by insufficient production of insulin. When type 2 diabetes is seen in the young, it is called maturity onset diabetes in the young (MODY). Management of diabetes is (1) Diet and Exercise (2) Oral hypoglycaemic drugs, e.g., sulfonylureas (glipizide, glimepiride, gliclazide), biguanides (metformin, pioglitazone), Voglibose, DPP-IV inhibitors (sitagliptin, vildagliptin, linagliptin-can be given in chronic kidney disease), and (Insulin-soluble or regular insulin, human mixtard insulin-30/70 or 50/50 and long-acting insulin-Lantus, glargine). A combination of two types of insulin is sometimes required. Secondary causes of diabetes are Thyrotoxicosis, Cushing's syndrome, Pheochromocytoma, Gigantism and Acromegaly, and iatrogenic-steroid treatment). Complications are Retinopathy, Coronary artery disease, Nephropathy, Neuropathy including peripheral neuritis and autonomic neuropathy, diabetic foot and large vessel (atherosclerosis) disease and Diabetic Ketoacidosis (DKA) which is a medical emergency.

Metabolic syndrome is characterised by abdominal obesity (abdominal girth > 40 inches in men and > 35 inches in female), insulin resistance (impaired glucose intolerance), hypertension and raised serum triglyceride (TG > 150 mg/dl). These patients are much more prone to develop coronary artery disease and stroke.

Cough

Causes:
- Common cold
- Infections of the respiratory tract
- Pneumonia
- Asthma
- Chronic obstructive pulmonary disease
- Pulmonary oedema
- Tuberculosis
- Bronchogenic carcinoma
- Left ventricular failure

- ACEs
- Foreign body
- Smoking

Tuberculosis

Tuberculosis (TB) is a chronic granulomatous infection commonly affecting the lungs and also lymph nodes, bones, intestine, meninges, urinary system and other organs. TB is caused by *Mycobacterium tuberculosis*. The bacillus, when it infects children, a primary focus forms which heals. The lesion is known Ghons focus. Lung TB is manifested by chronic low-grade fever, chronic cough persisting for more than 3 weeks, loss of weight and sometimes coughing out of fresh blood (haemoptysis). When the disease affects the lymph nodes (commonly cervical lymph nodes in children), they become enlarged and matted. Non-healing sinuses may be formed. When the bacillus affects the meninges, the condition is called meningitis. Miliary tuberculosis occurs in the lungs and is a serious condition.

Diagnosis in made by chest X-ray, Blood count (ESR is raised significantly) and sputum examination for AFB (acid fast bacillus). TB endemic area, Mantoux test is not of much significance.

Alert: *Any cough persisting for more than 3-4 weeks should be investigated for a cause which may sometimes be serious enough and will give the scope to treat before it is too late.*

Investigations: H/o duration, fever, drug history, Blood count, ESR, CRP, Sputum for AFB, Gram Stain, Chest X-ray (PA), HRCT-Scan thorax, MRI thorax, Bronchoscopy, CT-guided FNAC, ECG, Echocardiography. Biopsy of accessible lymph nodes may be useful. Culture of biopsy material may be done, but it requires 3-4 weeks to get the result. *QuantiFERON-TB Gold test for detecting* Mycobacterium tuberculosis *infection may be recommended.*

Pulmonary tuberculosis is treated with four drugs (INH, Rifampicin, Pyrazinamide and Ethambutol) for 2 months and followed by 3 drugs (INH, Rifampicin and Ethambutol) for 4 months. Pyridoxine should be given for entire 6 months to prevent peripheral neuropathy due to INH. Compliance is an important issue and should be addressed properly. DOTs therapy (Directly Observed Treatment) will ensure compliance and is observed in India. BCG vaccination is recommended for all new borne in India.

Delirium

Delirium, defined as an altered mental status with confusion, commonly occurs in older people and sometimes in patients staying for longer period in ICU (ICU psychosis). The precipitating causes, besides old age, are infection, sepsis, uremia and hepatic encephalopathy, lack of sleep, parkinsonism, antihistamines, morphine, datura poisoning, alcohol withdrawal, post-dialysis, thyrotoxicosis, high fever and others. The condition should be recognized when a person suddenly behaves abnormally in the setting of the ailments mentioned above. Treatment is primarily directed to the underlying condition, e.g., treatment of infection with antibiotics, intravenous fluid and symptomatic treatment with haloperidol, quetiapine, olanzapine or chlordiazepoxide and diazepam for alcohol withdrawal delirium.

Dementia

Dementia: Delirium is differentiated from dementia which is a chronic progressive brain disorder as opposed to delirium which is of sudden onset. Parkinsonism, Alzheimer's disease, HIV infection, Brain injury, Huntington's disease, Thiamine deficiency (Korsakoff's psychosis), and vascular diseases are some of the causes of dementia. Dementia patients have forgetfulness, inability to perform daily tasks, delirium, anxiety, depression and difficulty in concentrating etc. The diagnosis of dementia should be done only by exclusion of treatable causes and positive symptoms of the conditions. A battery of blood tests and imaging (CT-scan/MRI) may be required. Managing such patients is difficult for the care-givers since there is no specific curative treatment.

Visual disturbances

Visual disturbance is characterized by abnormal vision.

Causes: Blindness, Double-vision, Dimness of vision, Colour blindness, Pain, Shrinkage of field of vision, migraine-related visual disturbances

Blindness: A blind person does not have perception of vision. Causes: Glaucoma, Macular degeneration, Cataract, Optic neuritis, Retinitis pigmentosa, Corneal opacity. Risk factors: Diabetes, Hypertension, Vitamin A deficiency, Contact with chemicals. Recommended: Medical Students and doctors should learn Ophthalmoscopic examination as part of clinical evaluation of patients with diabetes, hypertension, neurological diseases, glaucoma, blindness, optic neuritis, medication toxicity (e.g., ethambutol, chloroquine).

Ptosis

Ptosis means drooping of the upper eye lid(s). Causes: third cranial nerve palsy, myasthenia gravis, external ophthalmoplegia, Horner syndrome (ptosis, miosis and anhidrosis of the face).

Diplopia

Diplopia stands for double vision which means that the person sees two overlapping images of a single image. Causes: Refractory error, Cataract, Stroke, Hyperthyroidism, Transient ischaemic attack (TIA), Brain tumour, Diabetes mellitus etc.

Nystagmus

Nystagmus means rapid, uncontrolled and repetitive movements of the eyes. Nystagmus may be congenital or acquired. Causes: Cerebellar tumour, Refractory error, stroke,

Multiple sclerosis etc. Nystagmus could be horizontal, vertical and rotatory.

Tremor

Tremor is usually manifested in the outstretched hand and fingers. Tremor can be fine or coarse.

Causes:

- Anxiety
- Familial
- Senile
- Thyrotoxicosis
- Parkinsonism
- Salbutamol
- Adrenaline
- Excess thyroxine supplementation

Anxious patients may show tremor due to over activity of the sympathetic nervous system, e.g., before examination and pass urine frequently.

Resting tremor, muscular rigidity (cog-wheel) and bradykinesia are features of Parkinsonism. Exophthalmos, tremor, tachycardia, anxiety, diarrhoea, loss of weight and insomnia are some of the features of thyrotoxicosis. Estimations of thyroid hormones (T3, T4, TSH, Free thyroxine, USG thyroid gland, Radio-isotope scanning ^{131}I, or ^{91}Tec, biopsy from thyroid nodule are required. MRI/ CT-scan brain is required if a pituitary cause is suspected. H/O intake of β-agonists (salbutamol, terbutaline, adrenaline) will give clue to diagnosis of tremor. Asthma/COPD patients take β-agonists.

Hypothyroid patients when given more thyroxine than required may show tremors.

Management: Treat the cause; withdraw the offending drug, Propranolol can control tremor, especially given to patients with thyrotoxicosis to control tremor till the anti-thyroid medications control the liberation of excess thyroxine.

Headache

Headache is pain in the head. It can be mild to severe, lasting hours to days, may be incapacitating, interfere with daily life, may be associated with vomiting (raised intracranial tension-meningitis, brain tumour, subarachnoid haemorrhage), intake of medications-indomethacin, isosorbide dinitrate, nifedipine. Deprivation of adequate sleep may cause headache. Mild headache can be managed by paracetamol, ibuprofens. Treat the cause.

Types:
- Migraine
- Tension headache
- Cluster headache
- Temporal arteritis

Treatment of Migraine: Paracetamol, NSAIDs, Sumatriptan, Ergotamine, Propranolol, Amitriptyline, Anxiolytics etc. Ergot preparations should be used carefully since they cause vasoconstriction and interfere with tissue perfusion of the coronary arteries and peripheral arteries.

The vertebral column consists of 33 vertebrae (cervical 7, C1-C7; thoracic 12, T1-T12, lumber 5, L1-L5; sacral 5, S1 -S5 and coccygeal 4,C1-C4). The first 24 vertebrae are distinct but the sacral and coccygeal vertebrae are fused. The vertebral foramen housing the spinal cord is situated at the centre of the vertebral body and the arch. The spinous process and the two each of the transverse and articular processes are situated behind the vertebral body.

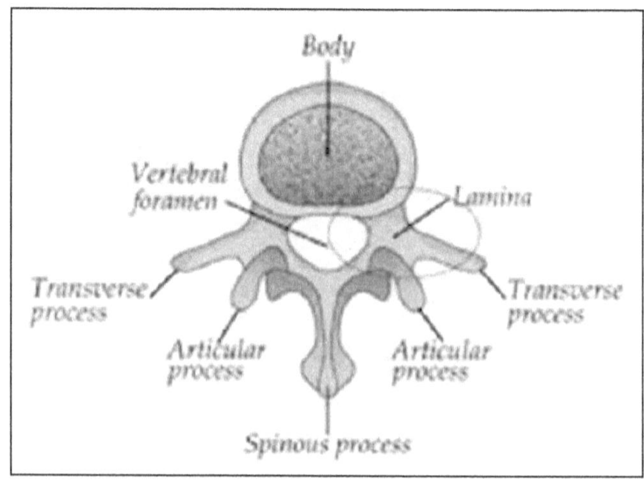

Figure: vertrebral column

Backache

Backache is a common medical presentation which creates troublesome and painful chronic illness. The condition is created by the muscles, joints, ligaments and bones. Backache is more common in adults. Sciatica causes pain the lumber region and radiates towards the legs. Sudden lifting of heavy weight may cause severe backache due to prolapse of the intervertebral disc.

Causes:
- Lumber spondylosis
- Prolapsed intervertebral disc
- Trauma
- Bony metastasis (carcinoma prostate, breast, lung)
- Osteoporosis
- Fracture and collapse of vertebral body
- Degenerative changes

Management: Blood sugar, CRP, PSA, X-ray LS spine (AP, Lateral), MRI etc. Complete rest in supine position in hard mattress will give relief from pain. Analgesics (Paracetamol) and anti-inflammatory drugs (Ibuprofen, diclofenac sodium etc.) and muscle relaxants, e.g., diazepam may be required to control pain. Opioids should be avoided for possible addiction. Cervical spondylitis is a relatively common problem in adults and elderly. Pain in the nape of the neck often radiates towards the arms in cervical spondylosis. Such patients often complain of vertigo which is aggravated by movement of the neck (positional vertigo). Cervical collar may help relief from pain by minimizing the movement of the cervical vertebrae so that the compression of the emerging from the spinal cord does not take place. These patients should be referred to Neurosurgeon/Orthopedic surgeon for evaluation and appropriate intervention.

Bone and Joint pain disorders

Bones form part of the skeleton that gives stability/mobility of the body, protects the internal organs, and the bone marrow (found inside the hip, thigh, ribs, vertebrae, skull bones and other) contain pluripotent stem cells that serves as precursor of the RBC, WBC and platelets that form part of the circulating blood. Osteoporosis and osteomalacia are bone diseases due to lack of intake of calcium and vitamin D. These patients are prone to bone fractures. Multiple myeloma is an important malignant bone disease. It causes severe bone pain and punched out lesions in the skull bones demonstrated by X-ray of the skull. Bence Jones Protein in urine and M pattern of immunoglobulin produced by the plasma cells in the plasma immunoelectrophoresis gives the diagnosis of Multiple myeloma. Anaemia is present as the excess immunoglobulin damages the kidneys. Treatment of the malignant bone diseases should be done by Oncologists. Ricket's affect children caused by deficiency of calcium and vitamin D.

Shoulder joint pain may be due to osteoarthritis. Gradual loss of movement of the shoulder joint leads to frozen shoulder. Pain killer and physiotherapy is generally all that is required. Surgery may be required in exceptional cases not responding to conservative treatment. Osteoarthritis of the knee exceptionally interferes with movement of the joint and the patient cannot walk without pain. This leads to physical disability. It occurs in adults of both sexes above the age of 40-50 and becomes worse with advancing age. Similarly hip joints are affected. Calcium, Vitamin D_3 and glucosamine/ chondroitin sulphate are used with limited joint improvement. Hip/knee replacement is undertaken and this gives excellent relief from pain and mobility becomes almost normal.

Gout

Gout is typically manifested by severe pain in the right big toe which becomes red, hot, swollen and painful. Normally in this condition Serum Uric Acid level is > 7 mg/dl, but may also be normal. Treatment is to give relieve of pain by using Colchicine, NSADs (e.g., indomethacin). Chronic gouty attacks can be prevented by allopurinol (100-300 mg daily after breakfast or febuxostat (40-80 mg daily).

Deep Venous Thrombosis (DVT): DVT means thrombosis of the deep veins particularly of the leg veins. Risk factors: prolonged immobilization, old age, long air travel, malignancy, polycythaemia, trauma, intravenous drug abuse, pregnancy, oral contraceptive. Management: keep the leg elevated, DVT stocking, injection of heparin. Complication: Pulmonary Thromboembolism (PE).

Cataract

Cataract is opacification of the lenses resulting in diminution of vision, partly or completely. The cataract formation is a slow process. Cataract occurs in the elderly. Diabetes and trauma are important causes of cataract. Recommended: all elderly patients coming for the first visit must be examined for cataract. Before contemplating for surgery, screening and management of diabetes, hypertension and dyslipidaemia is essential. Cataract is essentially removal of the opaque lenses and replacement by a plastic transparent lens. Types of cataract surgery are: Phacoemulsification, or phaco and extracapsular cataract surgery. Phaco requires a small incision at the margin of the cornea and a small ultrasound is inserted which dissolves the existing opaque lens. This is followed by insertion of a new lens. Computer-guided Femtosecond laser surgery is now very popular Cataract surgery.

Sleep disorders

Normal sleep: 7-9 hours (average 8 hours) of sleep at night is adequate for an adult.

Insomnia is defined as inability to fall asleep, continue to sleep and wakes up with tiredness. Becomes sleepy during daytime. Insomnia may lead to anxiety and depression. Chronic pain due to any cause may lead to insomnia. Too much sleep during daytime may cause insomnia. Excessive caffeine and alcohol intake contributes to insomnia. Diabetes mellitus and enlarged prostate cause increased frequency of urination so that the person has to get up from sleep for one or more times. It has been suggested, if possible, water drinking should be restricted at night.

Obstructive Sleep apnoea (OSA) is a sleep disorder caused by mechanical obstruction of entry of air into the lungs during sleep. As a result, oxygen does not enter the lungs and brain does not get sufficient oxygen (apnoea). When the person sleeps, the tongue falls back and the airway is obstructed. At that point, the person snores, the obstruction is opened and air enters the lungs (hypopnea). The apnoea and hypopnoe continues throughout the knight. The patient does not sleep and daytime somnolence interferes with daytime activity. OSA patients are prove to develop hypertension and diabetes. Hypothyroidism if present should be treated. Continuous Positive Airway Pressure (CPAP) is highly effective when taken at night during sleep. Compliance is an issue.

***Restlessleg syndrome* (RLS)** is a condition when immediately after few minutes of sleep, the patient wakes up to experience a peculiar type of leg pain and tries to get relief from pain by changing position of the leg(s). The patient also keeps the leg hanging beside the bed for relief of pain. Leg pain is relieved after walking for several minutes. The patient complains of daytime somnolence. Dopaminergic drugs, e.g., Levodopa + carbidopa combination, (used for the treatment of parkinsonism) and clonazepam are effective. The cause of RLS is not known. Pramipexole has been found to be highly effective when taken at 7-00 pm daily single dose. Since the anti-parkinsonian drugs are effective in this condition, it is presumed that RLS may be caused by a similar mechanism of action of the neurotransmitters.

Dyspnoea

Dyspnoea is defined as shortness of breath. In this condition the person becomes conscious of laboured breathing. Acute dyspnoea can be caused by diseases which included left ventricular failure (LVH) due to left sided valvular diseases (mitral stenosis, mitral incompetence, aortic stenosis) and hypertension. In left sided heart failure, sudden dyspnoea accompanied by cough. Many lung conditions e.g., bronchial asthma, chronic obstructive pulmonary disease (COPD), pulmonary oedema can cause dyspnoea. The other conditions are myocardial infarction, carbonic monoxide poisoning, severe anaemia, anaphylaxis etc.

Patient should be propped up, oxygen given, loop diuretics (furosemide) intravenously, morphine, slow lowering BP in patents with hypertension, positive inotropes etc. Right sided heart failure causes dyspnoea, pedal oedema, raised jugular venous pressure (JVP) and enlarged lover.

Dysphagia

Dysphagia is characterised as difficulty in swallowing. Dysphagia should be differentiated from the feeling that something is present in the throat in the absence of an organic cause and is known as Globus Hystericus. Dysphagia may be the presenting symptom of several serious diseases, e.g., carcinoma of the oesophagus, carcinoma of the stomach obstructing the lower end of the oesophagus. These cancers usually occur in adults and older people. The motility of the sphincter situated in the lower end of the oesophagus cannot relax in time to allow the food entering the stomach from the oesophagus-a condition known as Achalasia cardia. Oesophageal stricture may develop when strong acids, e.g., sulphuric or hydrochloric acid and alkalis, e.g., are ingested accidentally or trying to cause suicide. The stricture(s) cause mechanical obstruction for swallowing (dysphagia).

Enlargement of the liver

- Infective hepatitis
- Cirrhosis of Liver (early stage)
- Malaria
- Kala-azar
- Typhoid fever

- Congestive cardiac failure
- Hepatic carcinoma
- Amoebic liver abscess
- Alcoholic liver disease
- Non-alcoholic fatty liver

Enlargement of the spleen

- Malaria
- Kala-azar
- Infective hepatitis
- Typhoid fever
- Infectious mononucleosis
- Bacterial endocarditis
- Cirrhosis liver with portal hypertension
- Haemolytic anaemia
- Leukaemia
- Multiple myeloma
- Hodgkin's disease
- Myelofibrosis

Enlargements of liver and spleen

- Acute viral hepatitis
- Infectious mononucleosis
- Malaria
- Leishmaniasis
- Leukaemia
- Lymphoma

- Sickle cell anaemia
- Thalassemia
- Myelofibrosis
- Amyloidosis
- Systemic lupus erythematosus,
- Sarcoidosis
- **New-borns**: Thalassemia ; **Older children**: Malaria, Kala-azar, Typhoid fever, Sepsis

Common Skin Diseases

Scabies:

Scabies is caused by *Sarcopti scabei*. Scabies occurs in children more often than adults. It generally occurs in those living in over crowed places and having poor hygienic condition. Scabies is highly communicable and occurs in almost all family members at the same time. It spreads by sin-to-skin contact. Multiple papulo-vesicular rashes are seen in the webs of the fingers, typically, the ganitalia, buttocks and other parts of the body. The lesions are intensely itching and more so at night. Secondary bacterial infections may supervene. Sometimes, Scabies may precede the development of acute glomerulonephritis. Topical application of gamma benzene hexachloride, benzyl benzoate and ivermectin are highly effective. Secondary infections should be treated. All the family members should be treated simultaneously.

Leprosy:

Leprosy occurs predominantly in the tropics and subtropical areas. It is caused by *Mycobacterium leprae* which produces a granulomatous infection of the skin and affects the nerves. Prolonged contact with leprosy patient generally may transmit the infection to the uninfected individual. Leprosy may be classified into three broad types-lepromatous, tuberculoid and borderline. The lepromatous type is the most severe form of the disease and causes disfigurement of the face (leonine facies). The tuberculoid form manifests as anaesthetic depigmented/hypopigmented skin lesions. The nerves are thick and tender. It easy to see the thick tortuous great auricular neve and ulnar nerve in the ulnar groove. Rifampicin, clofazimine, dapsone are used or the treatment of Leprosy. Lepra reactions are sudden appearance of inflammatory reactions in patients with leprosy and requires specialised treatment.

Psoriasis:

Psoriasis Is a chronic recurrent inflammatory disorder involving skin, nails, and joints. Well defined, indurated erythematous plaques covered by adherent silvery scales are seen on the extensor aspect of elbow, knee and affects the joints and nails. The scales can be removed in layers (Grattage sign) and on removal; some punctuate bleeding spots appear (Auspitz's sign). Psoriasis is treated with immunosuppressive drugs, e.g., prednisolone, methotrexate, sulfapyrazone and hydroxychloroquine.

Acne:

Acne is also known as pimples. Pimples usually start appearing on the face in many boys and girls during puberty. Pimples may occur in other parts of the body e.g., back, neck etc. The hair follicles, when they are blocked by sebum, pimples are formed. Treatment of pimples include: eating less carbohydrate, washing face, keep the face clean and do not rub the pimples, otherwise healing is delayed with scarring which gives the face a bad appearance; medications are: local applications e.g., benzoyl peroxide, retinoic acid; orally tretinoin.

Boil and Carbuncle

Boil and Carbuncle are painful infections of the hair follicle under the surface of the skin. The follicles become inflamed and contains pus. The infecting organism is *Staphylococcus aureus.* When several boils coalesce together, it is called carbuncle. When a person gets repeated boils/caruncles, screening for diabetes should be done. Treatment with a suitable antibiotic, e.g., erythromycin/clarithromycin are effective. The accumulated pus should be drained aseptically.

Eczema

Eczema is caused by constant irritation of the skin. It may occur more frequently in atopic individuals. It may cause intense itching in night interfering with sleep. Treatment with low potency corticosteroid is required, but long-term treatment causes thinning of the skin.

Tinia

Tinea or ringworm is a fungal disease occurring in the groin of individuals who do not have proper hygienic habits. Teniasis can affect the nails. The lesions itch. Treatment is with 1) Salicylic acid (local application), 2) Oral drugs: Griseofulvin, Fluconazole, Terbinafine. Nail infections require long term treatment.

Urticaria

Urticaria are swollen, itchy, red lesions which fade after sometime. They occur in atopic individuals. In more severe cases, the eyes become swollen and may even cause laryngeal oedema and if not treated urgently the patient may die. Treatment with antiallergics, e.g., cetirizine, fexofenadine, loratadine, diphenhydramine is effective. Laryngeal oedema requires urgent treatment with adrenaline (0.3 ml IM). Steroids are frequently used along with antihistamines. Angioneurotic oedema and anaphylaxis are similar serve illness.

Retention of urine

Retention of urine means that urine is formed by the kidneys but it cannot be passed by the person voluntarily. The condition causes pain due to stretching of the bladder. Percussion of the lower abdomen will show dullness if the bladder is full. It should not be confused with anuria (no urine formed). The causes are:

- Enlarged prostate in elderly men
- Benign prostatic enlargement

- Bladder neck obstruction
- Bilateral obstruction of the urethra by renal stone
- Urethral obstruction
- Anticholinergic drugs in the elderly
- Neurogenic bladder (spinal cord injury, parkinsonism etc)
- Trauma to pelvis

Relief of obstruction:

- Encourage the patient to pass urine (standing may be useful)
- Hot and cold-water application over the lower abdomen
- Catheterisation
- Treatment of the cause of the obstruction

Nocturnal polyuria

Nocturnal polyuria is a condition where the individual has to get up at night for urination. Drinking too much water during the day particularly afternoon makes person to get up at night from sleep to urinate. Diuretics should be taken early morning. Enlargement of the prostate (males only), diabetes mellitus and urinary tract infection are some of the causes of nocturnal polyuria. Some children suffer from nocturnal enuresis commonly known as bed-wetting. Desmopressin acetate (DDAVP) reduces urine formation by the kidneys and an antidepressant (imipramine) have been successfully used for the control of nocturnal enuresis in children.

ECG GLOSSARY

A few ECGs are shown in this section. These are some of common ECG changes seen amongst patients seen at the ICU.

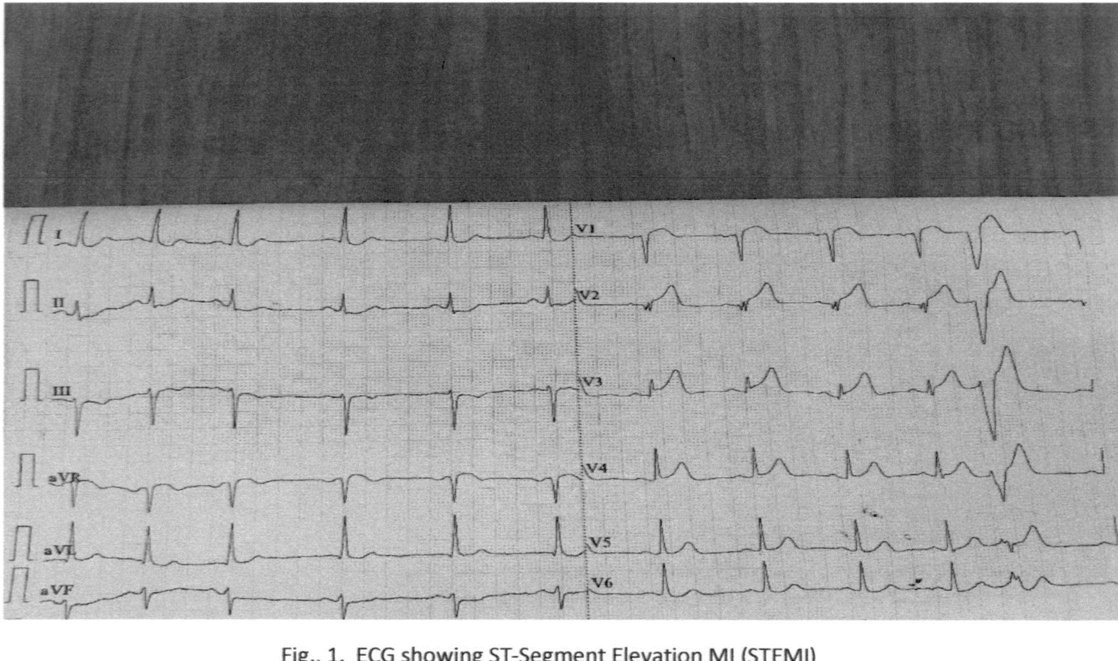

Fig., 1. ECG showing ST-Segment Elevation MI (STEMI)

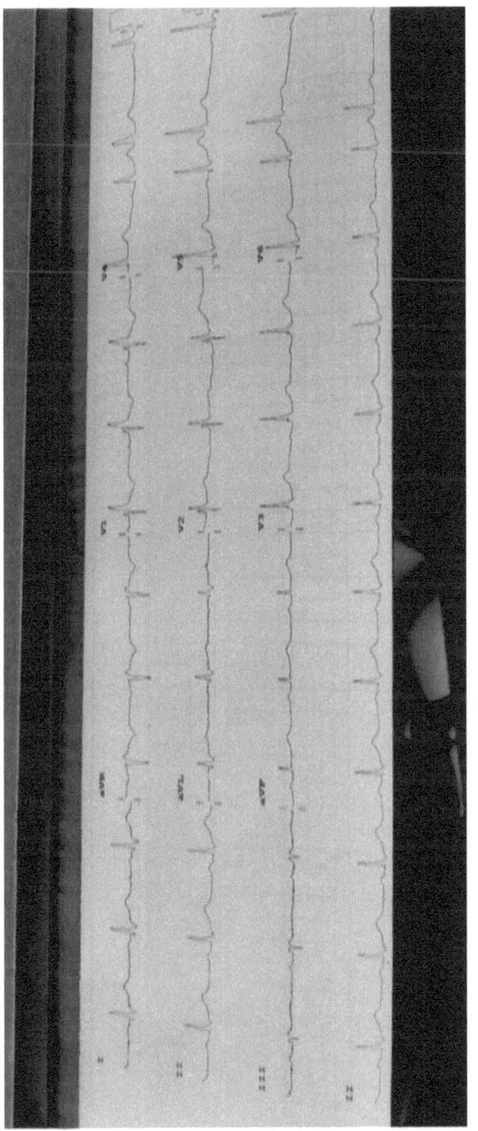

Fig., 2.ECG showing RBBB

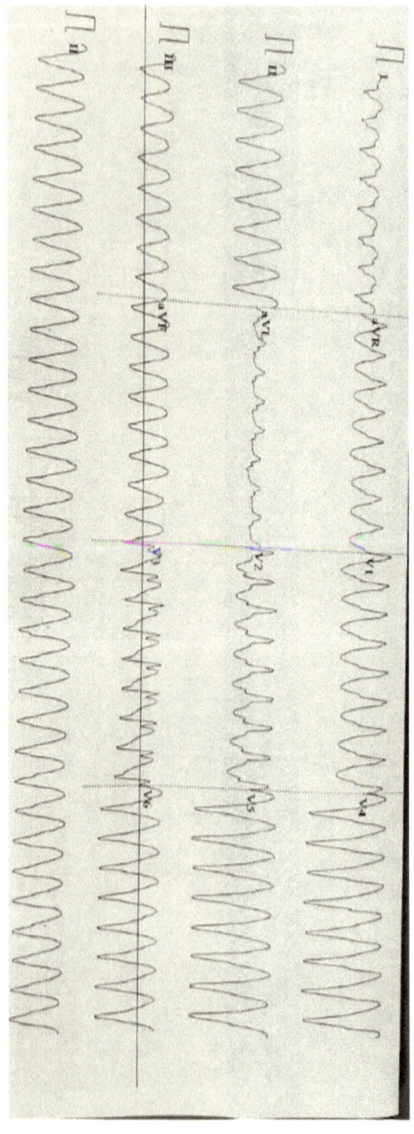

Fig., 3. ECG showing Ventricular Tachycardia

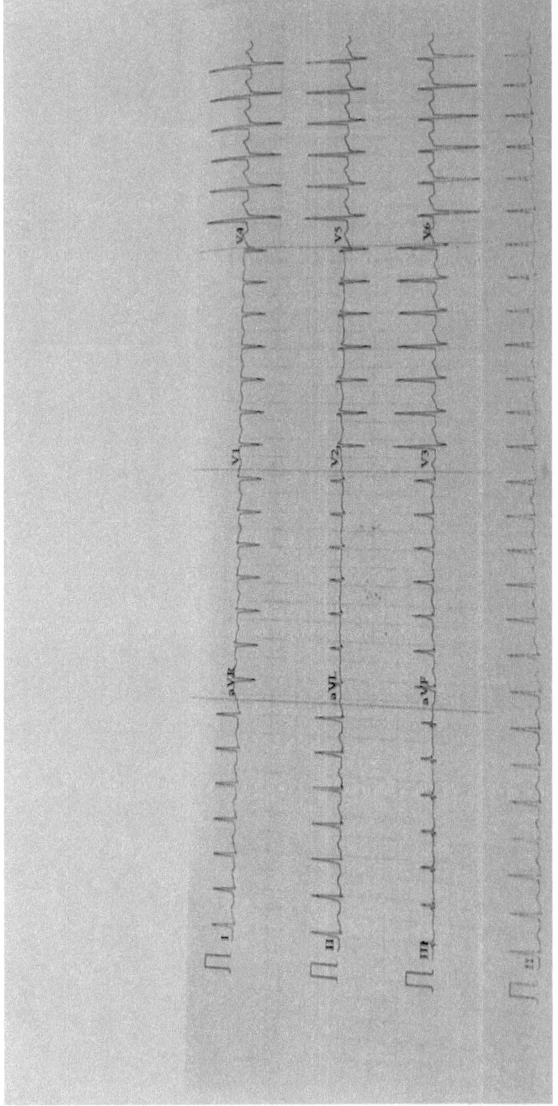

Fig., 4. ECG showing SVT

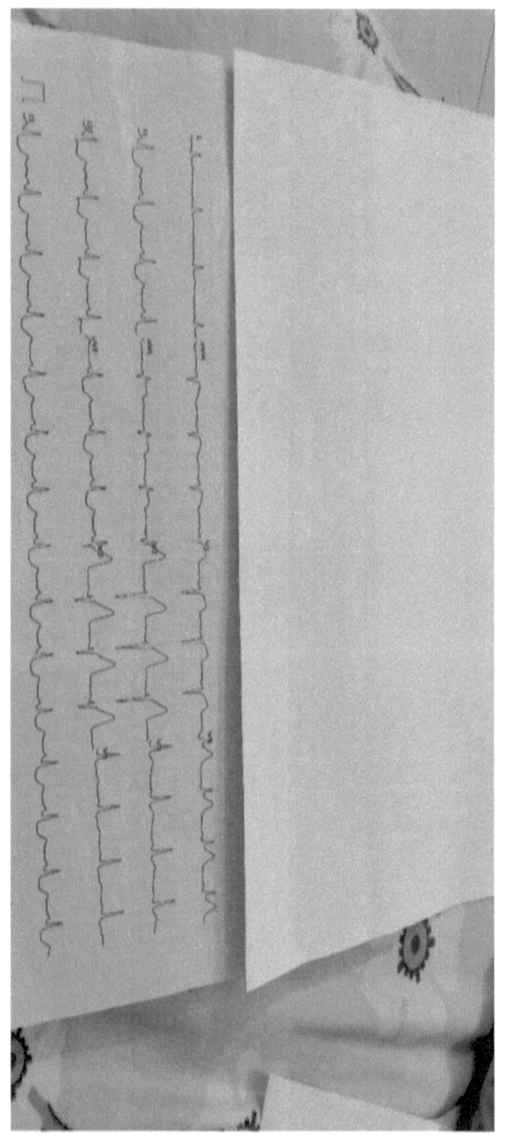

Fig., 5. ECG showing ST-Segment Elevation MI (STEMI)

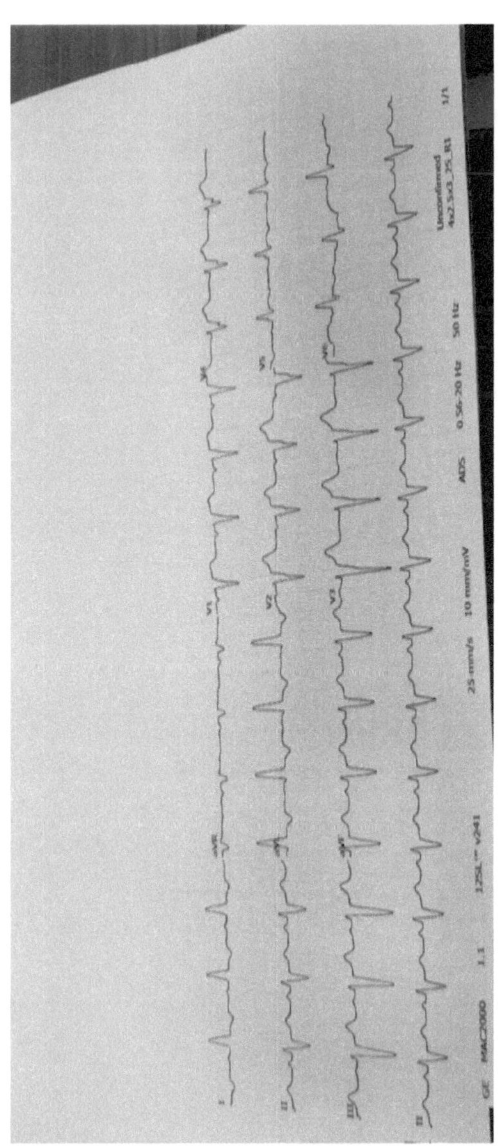

Fig.,6. ECG showing LBBB

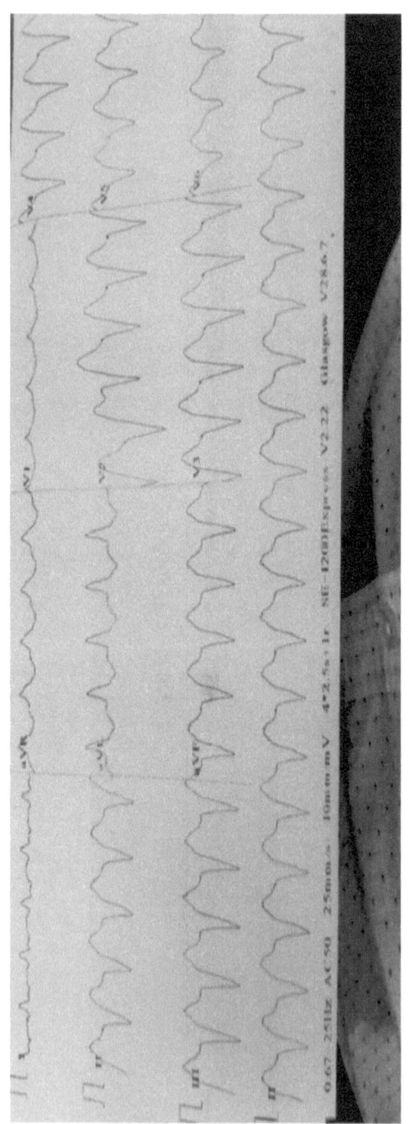

Fig., 7. ECG showing: Pace-maker rhythm